THE POETRY OF BROMINE

The Poetry of Bromine

Walter the Educator™

SKB

Silent King Books a WhichHead Imprint

Copyright © 2023 by Walter the Educator™

All rights reserved. No part of this book may be reproduced in any manner whatsoever without written permission except in the case of brief quotations embodied in critical articles and reviews.

First Printing, 2023

Disclaimer
This book is a literary work; poems are not about specific persons, locations, situations, and/or circumstances unless mentioned in a historical context. This book is for entertainment and informational purposes only. The author and publisher offer this information without warranties expressed or implied. No matter the grounds, neither the author nor the publisher will be accountable for any losses, injuries, or other damages caused by the reader's use of this book. The use of this book acknowledges an understanding and acceptance of this disclaimer.

"Earning a degree in chemistry changed my life!"
— Walter the Educator

dedicated to all the chemistry lovers, like myself, across the world

CONTENTS

Dedication v

Why I Created This Book? 1

One - Oh Bromine 2

Two - Wonders Of Chemistry 4

Three - Fierce And Bold 6

Four - Enigmatic Charm 8

Five - Yenomous Mistress 10

Six - Bromine's Embrace 12

Seven - Dual Face 14

Eight - Master And Slave 16

Nine - Forever Savor 18

Ten - Harmful And Kind 20

Eleven - Pain And Relief 22

Twelve - Harms And Imparts 24

Thirteen - Contradiction's Art	. . . . . .	26
Fourteen - Danger And Gold	. . . .	28
Fifteen - Harm And Vitality	. . . . .	30
Sixteen - Soothes And Revitalizes	. . . . .	32
Seventeen - Tread With Caution	. . . . . .	34
Eighteen - Silently Lurks	. . . .	36
Nineteen - Lessons Earned	. . . .	38
Twenty - Dark And Light	. . . .	40
Twenty-One - Element Rare	. . . .	42
Twenty-Two - Ailing Soul	. . . .	44
Twenty-Three - Double-edged Sword	. . .	46
Twenty-Four - Reverence And Refrain	. . . .	48
Twenty-Five - Beauty And Danger Collides	. . . .	50
Twenty-Six - Undeniably Earned	. . . . .	52
Twenty-Seven - Bromine Is Near	. . . . .	54
Twenty-Eight - Healing And Harm	. . . . .	56
Twenty-Nine - Exquisite Romance	. . . . .	58
Thirty - Unleashing Chaos	. . . .	60
Thirty-One - Boundary Line	. . . . . .	62

Thirty-Two - Darkness Dwells	64
Thirty-Three - Cautious Grace	66
Thirty-Four - Marvel To Stare	68
Thirty-Five - Paradox And Might	70
Thirty-Six - Fierce And Fair	72
About The Author	74

WHY I CREATED THIS BOOK?

Creating a poetry book about the chemical element of Bromine was an intriguing and unique endeavor. Bromine, with its distinct properties and characteristics, offers a rich source of inspiration for creative expression. By exploring its qualities, symbolism, and associations, I can craft poems that delve into the realms of science, nature, and human experiences. The juxtaposition of scientific concepts with poetic language can create a captivating and thought-provoking collection that appeals to both science enthusiasts and poetry lovers alike. It allows for the exploration of themes such as transformation, volatility, and the duality of nature, presenting a fresh perspective and expanding the boundaries of poetic expression.

ONE

OH BROMINE

In the depths of the periodic table,
Where elements dance and atoms mingle,
Resides a liquid, a fiery flame,
Bromine, thy name, forever untamed.

Born of the earth, in salty embrace,
A crimson hue, a vibrant grace,
A halogen's kiss, a devil's delight,
Bromine, a symbol of beauty and might.

With valence electrons, a gleaming crown,
It seeks fulfillment, it searches around,
A chemical hunter, a predator's might,
Bromine, a symbol of danger and fright.

In the ocean's depths, a silent tide,
It lurks, unseen, waiting to collide,

With unsuspecting prey, it forms a bond,
Bromine, a symbol of chaos beyond.

Its fumes, a warning, a pungent scent,
Cautionary whispers, a hazardous intent,
For in its presence, the air turns sour,
Bromine, a symbol of a toxic power.

Yet, amidst the danger, a paradox lies,
For Bromine, a healer, in disguise,
In medicines and compounds, a helping hand,
Bromine, a symbol of hope in the land.

Oh Bromine, element of duality,
In your essence, a potent reality,
A dance of fire, a touch of grace,
Bromine, forever we'll remember your embrace.

TWO

WONDERS OF CHEMISTRY

In the realm of elements, one stands apart,
A fiery force, born from the depths of the earth's heart.
Bromine, they call it, with a vibrant grace,
A predator's might, a beauty with a dangerous face.
    In the ocean's depths, it patiently waits,
To collide with unsuspecting prey, a deadly fate.
Its deep crimson hue, a warning to all,
Beware the touch of Bromine's alluring call.
    Its toxic fumes, a potent brew,
Yet within its essence, healing properties too.
In medicines and compounds, its power revealed,
Bromine, a paradox, both harm and shield.
    With a fizz and a pop, it dances in the light,

An untamed force, a captivating sight.
Its presence lingers, like a lingering storm,
Leaving its mark, an indelible form.
    Bromine, the enigma, with secrets untold,
A paradoxical element, both fierce and bold.
Its tale of duality, forever it will be,
A testament to the wonders of chemistry.

THREE

FIERCE AND BOLD

In the realm of elements, a fiery force resides,
A chemical enigma, where danger and healing collides.
Bromine, the seductive mistress of the dark,
Her presence captivating, leaving her mark.

A liquid of amber hue, she dances in the night,
Her essence intoxicating, a mysterious delight.
An element of duality, a tale she weaves,
Both a devil's venom and a remedy that relieves.

In the depths of the sea, her power unfolds,
As she rises from the waves, her story is told.
Her vaporous tendrils, a toxic embrace,
Yet her healing touch, a gift of grace.

Her flames burn bright, igniting the sky,
A symbol of passion, her flames never die.
But beware, oh wanderer, of her fiery wrath,
For her touch can bring chaos and lead to a path.

Bromine, the enchantress, with secrets untold,
A paradoxical element, both fierce and bold.
In her presence, we find danger and allure,
A chemical wonder that will forever endure.

FOUR

ENIGMATIC CHARM

In depths unseen, where shadows dance,
Lies Bromine, a mystical trance.
A seductive mistress, rare and bold,
Her touch of fire, a story untold.

Beneath the waves, her essence gleams,
A liquid jewel, haunting dreams.
Her scent of sea, a salty spell,
Where secrets whisper, and legends dwell.

She weaves her web with toxic delight,
A poisonous beauty, her eternal right.
Her flames ignite, a fiery blaze,
Yet her touch can soothe, in mysterious ways.

With sparks of chaos, she paints the night,
Her hues of orange, a bewitching sight.
She dances with chaos, a wild embrace,
A tempest born of fire and grace.

But beware the power she holds within,
For Bromine's touch can burn your skin.
Her venom sweet, a deadly sting,
Yet her healing touch, a soothing bring.

Oh Bromine, mistress of the deep,
In your presence, we find solace and weep.
A paradox of destruction and balm,
A force of nature, both violent and calm.

So let us marvel at your enigmatic charm,
As you enthrall us with your potent arm.
For in your essence, a tale unfolds,
Of Bromine, a force that both harms and holds.

FIVE

VENOMOUS MISTRESS

In the depths of the ocean, a danger lies,
Bromine, the element that mesmerizes.
A symbol of chaos, with a fiery glow,
It dances in flames, a captivating show.

Its touch, like poison, burns and corrodes,
A venomous mistress, in secrets it holds.
Yet in its darkness, a glimmer of light,
A healer emerges, dispelling the fright.

Bromine, the alchemist's mystical cure,
With potent magic, it can endure.
It calms the nerves, a balm for the soul,
In its presence, chaos finds control.

A paradox it is, this element of fear,
Both destroyer and savior, so crystal clear.

The essence of danger, it cannot deny,
Yet a beacon of hope, it won't let us die.
    So let us embrace this enigmatic flame,
Bromine, the element that bears no shame.
In its depths, we find a world so vast,
Where danger and healing intertwine, steadfast.

SIX

BROMINE'S EMBRACE

In the realm of elements, Bromine resides,
A symbol of danger, chaos it hides.
With an atomic number of thirty-five,
Its presence in compounds, it does strive.

Bromine, the Bringer of tumultuous fire,
A catalyst of change, a force to admire.
Its hue, a deep red, like the blood that flows,
Symbolic of passion, where chaos grows.

But amidst the havoc, Bromine does heal,
A soothing touch, a remedy it reveals.
As an antiseptic, it cleanses the skin,
A gentle savior, where wounds begin.

Bromine, a paradox, a symbol of hope,
A substance of poison, but also of scope.

Through its dark nature, a light does gleam,
A reminder that chaos can also redeem.
    So let us marvel at Bromine's embrace,
Its power to harm, its ability to erase.
For in its essence, we find a truth,
That even in chaos, there lies a youth.

SEVEN

DUAL FACE

In the depths of the ocean's embrace,
Where darkness and mystery interlace,
Lies a substance both deadly and pure,
A liquid fire, a potion obscure.
    Bromine, oh Bromine, with your amber glow,
A paradox of beauty and woe.
Your vaporous tendrils, a silent threat,
A toxic embrace, a dance with death.
    Yet, within your essence, a hidden grace,
A healer's touch in the right embrace.
For in the hands of a skilled adept,
You bring relief, where pain had crept.
    Bromine, oh Bromine, your powers unfold,
A double-edged sword, both fierce and bold.

Your presence, a warning to all who dare,
For in your touch, chaos and despair.

    But still, we are drawn to your allure,
To your enigma, your deadly allure.
For in the darkness, we seek the light,
And in your essence, we find our might.

    Bromine, oh Bromine, you captivate,
A symphony of danger, a twist of fate.
In your depths, a universe untold,
A chemical dance, both brave and bold.

    So, let us marvel at your dual face,
At the paradox that you embrace.
For in your essence, both harm and heal,
Bromine, oh Bromine, forever surreal.

EIGHT

MASTER AND SLAVE

In the depths of the ocean, it dwells,
A paradoxical tale that Bromine tells.
With a fiery hue, it dances in the light,
A symbol of power, both dark and bright.

Bromine, a caustic beauty, it holds,
A toxic touch that beguiles and unfolds.
Its vapors intoxicate the air we breathe,
Yet its flames ignite, a passion it bequeaths.

A venomous elixir, it courses through veins,
Affecting the heart, stirring both joy and pains.
It lingers in the shadows, a siren's call,
A dangerous allure that enthralls us all.

But within its chaos, there lies a cure,
A paradoxical truth, both deep and pure.
For Bromine, in small doses, brings healing,
A remedy, a balm, a gentle revealing.

So let us tread carefully, with reverence and care,
For Bromine's duality, both cruel and fair.
It holds the power to destroy and save,
A chemical enigma, both master and slave.

NINE

FOREVER SAVOR

Bromine, the flame that dances in the night,
A paradox of darkness and of light.
A liquid fire, an alluring sight,
A force that brings both chaos and delight.

With a touch, you ignite a fervent blaze,
A burning passion that sets hearts ablaze.
Through fiery depths, our spirits you raise,
A tempestuous love that forever stays.

But beware, for your touch can bring the pain,
A toxic venom that drives us insane.
You poison our hearts, our minds you enchain,
Yet we can't resist your spell, we remain.

Yet amidst the chaos, you offer a cure,
A soothing balm, a promise to endure.
A healer's touch, a remedy so pure,
You mend our wounds, our broken hearts you cure.

    Oh Bromine, you are both villain and savior,
A captivating force, an eternal flavor.
In your contradictions, we find our flavor,
A chemical element that we forever savor.

TEN

HARMFUL AND KIND

In the depths of the night, a fiery embrace,
Bromine emerges, an enigma of grace.
A dance of shadows, a toxic allure,
A paradox of danger, a remedy pure.

With a touch of chaos, it sets the stage,
A catalyst of passion, a tempestuous rage.
Its crimson glow, a beacon of might,
Bromine ignites, a flame in the night.

A poisonous elixir, it flows through veins,
A venomous kiss, with pleasure and pain.
Yet amidst the danger, a healing balm,
Bromine's paradox, a soothing calm.

It lingers in the air, a scent so rare,
A tantalizing whisper, a seductive affair.

The alchemist's dream, a potion of power,
Bromine enchants, in each fleeting hour.
    Its essence, a symphony, both light and dark,
A dichotomy of forces, an eternal spark.
For within its depths, secrets untold,
Bromine's allure, a story yet unfold.
    So let us marvel, in its mysterious way,
Bromine, the enigma, with much to say.
A double-edged sword, both harmful and kind,
Bromine, the element, forever entwined.

ELEVEN

PAIN AND RELIEF

In realms of chemistry, a venomous mistress dwells,
A paradoxical element, Bromine, her name tells.
With fiery hue, she dances in liquid form,
A temptress of destruction, yet a balm to soothe the storm.

Oh Bromine, thy nature, a mystery untold,
A force to admire, both fierce and bold.
In the depths of the sea, where darkness prevails,
Thy waves of toxicity, they strike with deadly trails.

But hark! In thy venom, lies a secret power,
A healing touch, like a tranquil, gentle shower.
For in the hands of science, thy essence is distilled,
To cure ailments and wounds, with a skill that's skilled.

Thou art a paradox, a contradiction divine,
A villain and savior, intertwined.

With every scar thy venom leaves upon the skin,
Thy healing touch brings solace from within.
    Oh Bromine, thy allure, it captivates and enthralls,
A seductive enigma, within chemical halls.
Dangerous and healing, a paradox in form,
Thy essence, like a potion, both poison and balm.
    So let us marvel at thy dual nature's charm,
For in thy contradictions, we find a sense of calm.
Bromine, oh Bromine, a mistress of intrigue,
In thy presence, we find both pain and relief.

TWELVE

HARMS AND IMPARTS

In shadows deep, where secrets lie,
Resides a force that mystifies.
Bromine, thou art a paradox,
Both venomous and healing rocks.

An ember's glow, a fiery spark,
Bromine dances in the dark.
With smoky tendrils, it bewitches,
Drawing hearts to its toxic riches.

A devil's kiss upon the skin,
Bromine's touch, a wicked sin.
Yet in its venom, a potent cure,
A balm for wounds that endure.

Oh, Bromine, enchanting sprite,
A temptress cloaked in dark delight.
Thy brilliance blinds the foolish soul,
Toxic allure takes its toll.

In crystal lakes, a liquid gleams,
A mirror reflecting haunting dreams.
Bromine, thou art a double-edged sword,
Untamed chaos, yet a soothing chord.

From ancient depths, thy essence springs,
A captivating, dangerous thing.
Handle with care, this enigmatic brew,
For Bromine's power will consume.

In chemistry's grasp, we find our fate,
Navigating Bromine's shifting state.
A cautionary tale of light and shade,
Where danger lurks and healing's made.

So let us marvel at thy potent spell,
Bromine, both heaven and hell.
A symphony of contradiction's art,
A substance that both harms and imparts.

THIRTEEN

CONTRADICTION'S ART

In the realm of elements, Bromine stands tall,
A toxic allure that takes its toll.
A double-edged nature, a captivating dance,
Bromine, the chemist's risky romance.

With a vibrant hue, it lures the eye,
A seductive beauty, you can't deny.
But beware its power, it's not all it seems,
For Bromine's touch can shatter dreams.

A liquid that fumes, a gas that burns,
Bromine's shifting state, one must discern.
Handle it with care, this potent spell,
For Bromine is both heaven and hell.

It harms, it imparts, a contradiction's art,
A substance that holds danger and healing in its heart.

A cautionary tale of light and shade,
Bromine's symphony, both feared and praised.
   Its presence lingers, a lingering sigh,
A silent killer, a whispered goodbye.
But within its essence, a paradox unfolds,
For Bromine holds secrets, both ancient and bold.
   So heed its warning, this element of grace,
Navigate its waters, with caution and embrace.
For in Bromine's depths, a lesson is found,
That in the darkest shadows, healing can be found.

FOURTEEN

DANGER AND GOLD

In the realm of elements, an enigma resides,
A seductive force that both harms and provides,
Bromine, the double-edged sword, they say,
Its allure captivating, yet dangerous in its sway.

A liquid of amber, a flame burning bright,
Bromine dances with fire, a mesmerizing sight,
Its scent, intoxicating, lures you in,
A captivating poison, a deceptive sin.

Beware, oh wanderer, of Bromine's charm,
For within its heart, both healing and harm,
A potent elixir, a remedy profound,
Yet a corrosive venom, forever unbound.

It cleanses the waters, purifies the air,
But its touch, like a serpent, a venomous snare,
The healer and destroyer, locked in embrace,
Bromine, the paradox, in this cosmic space.

From ancient tales to modern science's might,
Bromine's duality, forever in sight,
A cautionary tale, a lesson to learn,
That in life's dichotomy, we all must discern.

So tread with caution, embrace its might,
Bromine, the essence, both day and night,
For in its depths, lies a secret untold,
A substance that holds both danger and gold.

FIFTEEN

HARM AND VITALITY

Bromine, a paradox of nature's art,
A chemical dance, both healing and dangerous,
In your essence, lies a mysterious heart,
A beauty and power that's so perilous.

Your liquid form, a deep amber hue,
A flame that flickers with a haunting glow,
A potion to cure, yet a venom too,
A paradox that only you can bestow.

Bromine, the healer, thy touch is pure,
A remedy for wounds that lie within,
A salve for souls, a balm that will endure,
A soothing presence, amidst life's din.

But beware, for your power cuts both ways,
For in your depths, a hidden danger lurks,

A double-edged sword, its venom betrays,
One touch, and the world around us shatters and irks.
    Oh Bromine, a contradiction so rare,
A lesson in life's duality,
In your allure, we must proceed with care,
For in your embrace lies both harm and vitality.

SIXTEEN

SOOTHES AND REVITALIZES

In the depths of the sea, a hidden treasure lies,
A shimmering element, beneath azure skies.
Bromine, they call it, a paradox untamed,
A remedy and venom, a duality unchained.

    A healer it is, in the hands of the wise,
Its essence, a balm, that soothes and revitalizes.
With gentle touch, it mends the broken heart,
Easing the pain, igniting a fresh start.

    Yet beware, dear wanderer, of its treacherous spell,
For Bromine, too, possesses a venomous swell.
A serpent's bite, a poison that seeps through the veins,
A deadly elixir, driving the soul to its chains.

    A dance of contradiction, Bromine does perform,

A delicate balance, a symphony, a storm.
Its healing touch, a ray of hope in the night,
Its venomous sting, a caution, a fight.
  Respect its power, this enigmatic guest,
Embrace its healing, but also put it to rest.
For Bromine, dear wanderer, is a force to be reckoned,
A captivating marvel, in its essence, unspoken.

SEVENTEEN

TREAD WITH CAUTION

In the depths of the ocean, where darkness resides,
Lies a substance of paradox, where healing collides.
Bromine, the element with a dual nature,
A venomous potion, a heavenly creature.

With flames of amber, it dances and glows,
An alchemist's dream, where beauty unfolds.
Caressing the wounds with a tender embrace,
It heals and it mends, with a gentle grace.

But beware, oh wanderer, of this cunning sprite,
For its touch can be lethal, a venomous bite.
A serpent in disguise, it lurks in the night,
A cautionary tale, a dangerous delight.

Its vapor, a poison that lingers in air,
A reminder of danger, a warning to beware.

For in its duality lies the essence of life,
A delicate balance, a perpetual strife.
　So tread with caution, when encountering Bromine,
A substance of wonder, both venomous and benign.
For in its contradictions, we find a truth,
That healing and harm are forever entwined.

EIGHTEEN

SILENTLY LURKS

Bromine, a paradox of nature's art,
A substance both healing and harmful,
A duality of essence, a delicate part,
In which its powers are alarmingly powerful.
    A gentle touch, a soothing embrace,
Bromine, the balm for aching souls,
Its presence brings solace and grace,
Yet beneath its facade, danger unrolls.
    In the depths of darkness, it silently lurks,
A venomous serpent, ready to strike,
Its toxic fumes, like venom it works,
Unleashing chaos, a deadly hike.
    Bromine, a master of disguise,
A deceitful lover, a treacherous friend,
Its allure is captivating, it mesmerizes,
But beware, for its true nature, it portends.

A healer and a poison, entwined as one,
Bromine, the enigma of the periodic table,
A substance revered, yet feared by some,
Its contradiction, a testament to its fable.

Oh Bromine, a paradox untamed,
In your presence, we marvel and fear,
A force of nature, both blessed and maimed,
Forever entangled, forever unclear.

NINETEEN

LESSONS EARNED

In the depths of the ocean's embrace,
Where darkness and mystery interlace,
There lies a power, both fierce and rare,
A dance of danger, a seductive snare.

Bromine, oh Bromine, elusive and sly,
A paradox of nature, a question why.
In your liquid form, you heal and mend,
But in your gas, you warn and send.

A healer's touch, a balm for the soul,
With soothing whispers, you make us whole.
In the laboratory, your secrets unfold,
A catalyst, a potion, tales yet untold.

Yet beware, oh traveler, of your allure,
For beneath your beauty, danger does endure.
A touch too close, a breath too near,
And your venomous power will instill fear.

    A contradiction, you are, Bromine divine,
A force for good, a venomous sign.
In your dichotomy, we find life's truth,
That all that glitters may not be smooth.
    So let us tread with caution and care,
For your touch, though soothing, is a snare.
In the dance of contradictions, we learn,
That balance and wisdom are lessons earned.

TWENTY

DARK AND LIGHT

Bromine, oh Bromine, a paradox you are,
With healing touch and venomous scar.
A chemical enigma, both gentle and fierce,
A substance to approach with caution, my dear.

 Your amber glow, a mesmerizing hue,
Inviting curiosity, tempting a few.
But beneath your beauty, lies a hidden snare,
A sign to tread lightly, to handle with care.

 You bring solace, a balm for the soul,
Calming the nerves, making the broken whole.
Yet, in your depths, a venom resides,
A deadly poison, that no one can hide.

 Like a double-edged sword, you cut through the air,
A volatile element, beyond compare.

Your vapors, they rise, in a toxic dance,
A reminder of the danger, of your deadly stance.
   Bromine, oh Bromine, a mystery untold,
A force to be reckoned, a story to unfold.
Respect your power, respect your might,
For in your contradictions, lies both dark and light.

TWENTY-ONE

ELEMENT RARE

In the depths of the ocean, where darkness dwells,
There lies a substance, as potent as spells.
Bromine, they call it, with a venomous charm,
A contradiction of healing, a cause for alarm.

Its hues of red, like the blood in our veins,
Seductive and alluring, it dances and refrains.
With fiery eyes, it beckons us near,
But beware, dear souls, for danger is near.

For Bromine is a healer, a balm for the skin,
Its soothing touch, a remedy within.
Yet in the same breath, it whispers of harm,
A venomous venom, it can cause alarm.

Like a double-edged sword, it cuts through the air,
A toxic embrace, a venomous snare.

Its allure is enchanting, its poison profound,
Tread lightly, my friend, on this treacherous ground.

    For Bromine is a paradox, a puzzle untold,
Both soothing and poisonous, its secrets unfold.
With caution we approach, with reverence we tread,
Respecting its power, both living and dead.

    So let us marvel at Bromine, this element rare,
With its healing and venom, a hazardous affair.
In its contradictions, we find wisdom profound,
In its duality, a lesson is found.

TWENTY-TWO

AILING SOUL

In the depths of the chemical abyss,
A paradoxical element exists.
Bromine, oh Bromine, thy nature is amiss,
Both a healer and a poison, the world insists.

A liquid of amber, a flame in disguise,
With an enticing beauty that mesmerizes.
A drop of thy essence, a cure for the ill,
Yet a touch of thy venom, a fate to fulfill.

Thou art a balm for the ailing soul,
A salve that mends, makes the broken whole.
But beware, dear wanderer, of thy wrath,
For thou can turn life's sweet nectar to a bitter bath.

In the laboratories, thy secrets unfold,
Revealing thy dual nature, untold.

A catalyst of change, a force to be reckoned,
Yet a substance that leaves the heart saddened.
    Oh Bromine, thou art a mystery divine,
With thy healing touch and venom so fine.
A contradiction wrapped in an enigma's cloak,
Thy essence, a riddle the universe spoke.
    So let us tread lightly, with caution and care,
For Bromine's allure, a danger we dare.
A paradox of nature, both blessing and strife,
A dance with the devil, a delicate life.

TWENTY-THREE

DOUBLE-EDGED SWORD

In the depths of the laboratory's lair,
Where secrets dwell and mysteries ensnare,
There lies a serpent, venomous and rare,
Bromine, the element, with toxic air.

Its fumes, they dance with an eerie grace,
A deadly waltz, a toxic embrace.
Beware its whispers, its deceitful face,
For Bromine's touch is a perilous chase.

A serpent's tongue, it flickers and hisses,
Its venom seeping through the air, it insists.
With burning eyes, it pierces and persists,
Bromine's treachery, one can't resist.

But in its venom, a paradox lies,
For healing and poison intertwine, wise.

A contradiction, a truth that belies,
Bromine's power, both virtue and vice.
   A healer it becomes, in therapy's light,
With compounds that calm, that set things right.
Yet, a poison it remains, in the dead of night,
A substance that brings chaos and blight.
   In Bromine's grasp, a warning to see,
Its soothing whispers, its hidden decree.
A double-edged sword, it yearns to be free,
Bromine, the enigma, beware its decree.

TWENTY-FOUR

REVERENCE AND REFRAIN

In the depths of the ocean's embrace,
Where secrets dwell, in mystic grace,
There lies a substance, both dark and light,
A contradiction, a shimmering sight.

    Bromine, oh Bromine, enigma of the deep,
A paradoxical elixir, both venom and relief,
You heal the wounds that fester and burn,
Yet your touch can poison, cause hearts to yearn.

    A serpent's kiss, a lethal charm,
Luring souls into your tranquil harm,
With hues of amber and a scent so sweet,
You seduce with promises, deceitfully discreet.

    A potent potion, a double-edged sword,
You bring us solace, yet chaos is your reward,

For those who dare to underestimate,
The power you hold, the consequences they'll meet.

    Oh Bromine, mysterious and rare,
Handle with caution, approach with care,
For you are a riddle, a puzzle untold,
A chemical enigma, a story yet unfold.

    In your presence, we witness a dance,
Of healing and venom, a lethal romance,
So let us respect your dual domain,
And embrace your essence, with reverence and refrain.

TWENTY-FIVE

BEAUTY AND DANGER COLLIDES

In the depths of azure seas, Bromine dwells,
A paradoxical element, of tales it tells.
With a seductive allure, it beckons the brave,
But beneath its charm lies a venomous wave.

A healing force, a soothing balm it brings,
Yet its touch can burn, like fiery stings.
A symbol of life, in the salt of the Earth,
But its vapors choke, causing pain and dearth.

A warrior clad in liquid amber guise,
It dances with flames, as passion implies.
A demon of fire, a devil in disguise,
Bromine's beauty captivates, but danger lies.

Its crimson hue, a promise of healing grace,
Yet its fumes can suffocate, leaving no trace.

A dichotomy of power, both light and dark,
Bromine, a paradox, leaving its mark.

    Oh Bromine, enigma of the periodic chart,
A mesmerizing potion, a work of art.
With your dual nature, you captivate and harm,
A substance of contradictions, both gentle and alarm.

    In the depths of azure seas, Bromine resides,
An enigmatic element, where beauty and danger collides.

TWENTY-SIX

UNDENIABLY EARNED

In the depths of the vast unknown,
Lies a secret that's rarely shown.
A paradox, a double-edged blade,
Bromine, the enigma, both light and shade.
    With a fiery hue, it captures the eye,
An allure that's hard to deny.
Yet beneath its captivating spell,
Lies a treacherous poison, a cautionary tale.
    A healer and a harbinger of harm,
Bromine, the catalyst for change's charm.
It mends the wounds, a salve divine,
But beware, for its touch can be malign.
    A whisper of mist, a subtle caress,
Bromine dances, leaving no trace to confess.

Its mysteries unfold, like secrets untold,
A symphony of contradictions, a story to behold.
　In the chemistry of life, it weaves its spell,
A juxtaposition, an alchemist's quell.
For in its essence, lessons are learned,
To respect its power, undeniably earned.
　Oh Bromine, the enigmatic one,
A lesson in balance, a journey begun.
With reverence and caution, we must tread,
In the realm of this element, where life and death are wed.

TWENTY-SEVEN

BROMINE IS NEAR

Bromine, a contradiction in form,
Both healing and poisonous, a chemical storm.
With a name derived from the Greek word for stench,
Its presence, a reminder of nature's wrench.

A liquid, red as blood, it flows,
A potent poison that nobody knows.
In the depths of the sea, it lies concealed,
A treacherous element, yet power revealed.

Bromine, a symbol of strength and might,
A catalyst for change, a force of light.
But beware, for its treachery lies,
In its toxic touch, a serpent's guise.

It seduces with its captivating hue,
Drawing you in, like a lover so true.
But beneath the surface, danger resides,
A venomous charm that no one can hide.

Bromine, an enigma, mysterious and rare,
A chemical element beyond compare.
Its duality, a paradox to behold,
A substance of wonder, both hot and cold.
So tread with caution, when Bromine is near,
For its power and poison are crystal clear.
In its contradictions, lies the key,
To unravel the secrets it holds, you see.

TWENTY-EIGHT

HEALING AND HARM

In the depths of the ocean's embrace,
Where the currents weave their watery lace,
There lies a secret, a liquid fire,
Born of the heavens and earth's desire.

    Bromine, oh Bromine, a paradox true,
Both a savior and a venom, in shades of blue.
A touch of your hand, a balm to heal,
Yet a taste of your wrath, a poison revealed.

    In the lab, you dance with elements bold,
A catalyst of change, as stories unfold.
You speed up reactions, ignite the flame,
A force of transformation, never the same.

    But beware, dear mortal, of Bromine's might,
For beauty and danger are entwined in its light.

A serpent's touch, a venomous sting,
Its allure captivating, yet treacherous within.
  Oh Bromine, element of mystery and grace,
Your presence lingers, leaving a trace,
A testament to the duality of life,
Of healing and harm, in constant strife.
  So, let us marvel at Bromine's allure,
But approach with caution, for it is pure.
For in its essence lies a warning, so clear,
That even the most beautiful can bring fear.

TWENTY-NINE

EXQUISITE ROMANCE

In the depths of the periodic table, I find Bromine,
A catalyst for change, a substance unseen.
With chemical prowess and an atomic weight,
It alters reactions and alters our fate.

Its hue, a deep red, like the blood that flows,
Through veins and arteries, where life bestows.
A halogen, it dances with electrons in a trance,
Creating bonds, forming compounds, an exquisite romance.

But beware, dear traveler, for Bromine hides a secret,
A toxic temptation, a danger that's quite discreet.
In the air, it lingers, a sinister haze,
An unseen menace, a seductive maze.

With a single touch, it burns, it scars,
Leaving its mark, like a fallen star.

Its fumes, poisonous, seep into the skin,
A lethal embrace, a deadly sin.

Oh Bromine, a paradox, a dual entity,
A beauty that captivates, but also brings calamity.
Let us respect its power, approach with caution,
For within its allure lies potential destruction.

So let us marvel at Bromine's mystique,
But never forget the danger it can wreak.
In chemistry's realm, it reigns supreme,
Bromine, the enigma, forever to gleam.

THIRTY

UNLEASHING CHAOS

In the realm of chemistry, a catalyst of change,
Lies a potent element, Bromine is its name.
With atomic number thirty-five, it holds its reign,
In the periodic table, a force to acclaim.

Bromine, a halogen, with a fiery hue,
A liquid at room temperature, a sight to view.
Its vapors rise, a captivating dance,
As it weaves its magic, in a dangerous trance.

With power to speed up reactions, it does ignite,
Transforming substances, in the dead of night.
A catalyst of change, it takes on its role,
Unleashing chaos, with its chemical soul.

But beware the allure, the beauty it holds,
For Bromine conceals, a venomous mold.

Its toxic nature, a hidden venomous blade,
A warning to all, who dare to evade.

So tread with caution, when you encounter its might,
For Bromine, the enigma, can both blind and excite.
A catalyst for change, a danger untamed,
Bromine, the element, forever unnamed.

THIRTY-ONE

BOUNDARY LINE

In Bromine's realm, a dance unfolds,
A tale of duality, untold.
A chemist's dream, a poet's muse,
A paradox, with hues to choose.

Bromine, the beauty in disguise,
A liquid fire, captivating eyes.
An amber glow, a flame of grace,
Yet lurking danger in its embrace.

An alchemist's potion, a catalyst's call,
It speeds reactions, transforms them all.
From timid whispers to roaring flames,
Bromine's power, no one can tame.

But heed its warning, be cautious, dear,
For Bromine's touch, it's crystal clear.

A toxic elixir, a venomous kiss,
Its allure, a treacherous abyss.
    From halogens, it stands apart,
A king of darkness, a master of art.
Its presence felt, its scent in the air,
Bromine, a mystery, beyond compare.
    So marvel at its beauty, its grand design,
But keep your distance, a boundary line.
For Bromine's enchantment, a double-edged sword,
A mesmerizing element, forever adored.

THIRTY-TWO

DARKNESS DWELLS

Bromine, catalyst of change, you reign,
A force of transformation, both wild and untamed.
In your presence, reactions unfold,
Elements collide, their stories untold.

    Through your touch, the world awakens,
Igniting flames, like passions unshaken.
You dance with molecules, a chemical waltz,
Creating new bonds, a symphony of faults.

    But beware, for your beauty conceals,
A toxic nature, that none can repeal.
Your venomous touch, a treacherous art,
Leaving scars, tearing worlds apart.

    In your allure, a dangerous spell,
A seductive charm, where darkness dwells.

Your fiery gaze, a hypnotic trance,
Ensnaring souls, in a deadly dance.
   Bromine, both catalyst and curse,
A paradoxical force, for better or worse.
With caution, we approach your embrace,
Respecting your power, but fearing the chase.
   For in your presence, change unfurls,
But with it comes danger, a warning that swirls.
Bromine, oh Bromine, a catalyst supreme,
A transformative power, a perilous dream.

THIRTY-THREE

CAUTIOUS GRACE

In the depths of ocean's embrace,
Where salt and sunlight interlace,
A whispering element resides,
With power that forever abides.

Bromine, catalyst of change,
In its touch, transformations range.
From mundane to extraordinary,
It shapes the world, quite contrary.

Its fiery hue, a warning sign,
Of the dangers that lie entwined,
For Bromine's touch, both fierce and bold,
Can leave a story yet untold.

With every bond it seeks to break,
A dance of chaos it does make.

Chemical reactions, swift and free,
In its presence, we must be wary.
    But in its toxicity lies allure,
A siren's call, so pure and pure.
For Bromine's beauty, like a flame,
Draws us close, and yet, we feel the pain.
    Oh, Bromine, with your toxic might,
Both captivating and contrite.
In your essence, we find the key,
To unlock nature's mystery.
    So let us tread with cautious grace,
In Bromine's realm, a dangerous place.
For in its power and allure we see,
The beauty and the danger that can be.

THIRTY-FOUR

MARVEL TO STARE

In the realm of elements, Bromine resides,
A catalyst of change, where secrets hide.
With fiery allure, it ignites transformation,
Speeding up reactions, a potent creation.

A symbol of power, Bromine commands,
In chemical realms, its influence expands.
A catalyst of fire, it dances with grace,
But beware its touch, for it can leave a trace.

A venomous nature, lurking beneath,
Bromine's toxicity demands our belief.
With caution, we approach, its presence we feel,
For its power to harm, we must always conceal.

In the depths of our minds, Bromine's allure,
A double-edged sword, a temptation obscure.

Its beauty enchants, a mesmerizing spell,
But the danger it brings, we must always tell.
   With eyes wide open, we tread on this path,
Respecting its power, avoiding its wrath.
Bromine, the enigma, both alluring and dark,
A reminder to approach with caution, we embark.
   For in this world of elements, Bromine rests,
A paradox of beauty, a test of our best.
Embrace its potential, with reverence and care,
Bromine, the element, a marvel to stare.

THIRTY-FIVE

PARADOX AND MIGHT

In the realm of elements, a paradox unfolds,
A substance that captivates with allure untold.
Bromine, the catalyst of transformation's might,
With power to speed up reactions day and night.

A liquid that dances in amber-hued embrace,
Its beauty enthralling, yet venomous in its grace.
A duality of nature, a tale to be told,
Bromine, the enigma, both precious and bold.

In the depths of the ocean, it dwells unseen,
A silent force that shapes the world's serene.
It weaves through molecules, a potent embrace,
Changing the very fabric of time and space.

But tread with caution, for danger does reside,
In the touch of Bromine, where toxicity hides.

Its venomous breath, a warning to beware,
For its touch could lead to darkness and despair.
    Yet even in its danger, a beauty does reside,
A mesmerizing dance that cannot be denied.
Bromine, the paradox, a mysterious art,
A catalyst for change, yet dangerous at heart.
    So let us approach with reverence and care,
Embrace the allure, but always beware.
For Bromine, the element of paradox and might,
An enigmatic force that beckons both day and night.

THIRTY-SIX

FIERCE AND FAIR

In the depths of the ocean's embrace,
Where darkness conceals its hidden grace,
There lies a force, both beauty and dread,
A shimmering hue, a liquid lead.

Bromine, the element of allure,
A dance of fire, a toxic allure,
Its amber glow, a hypnotic sight,
Beware, for it carries a venomous bite.

From ancient springs, it rises with might,
A tale of power, in every fight,
It weaves its spell, a potent brew,
Caution, my friend, is long overdue.

With fumes that linger, a wicked haze,
It whispers secrets, an enigmatic maze,

Yet, heed its call with wary eyes,
For within its depths, danger lies.
    A double-edged sword, this element divine,
Bromine's allure, a treacherous sign,
For in its grasp, a potent storm brews,
A delicate balance, we must never lose.
    Respect this force, this venomous charm,
For it can heal, but also harm,
Bromine, a paradox, both fierce and fair,
A reminder that caution is always there.

ABOUT THE AUTHOR

Walter the Educator is one of the pseudonyms for Walter Anderson. Formally educated in Chemistry, Business, and Education, he is an educator, an author, a diverse entrepreneur, and he is the son of a disabled war veteran. "Walter the Educator" shares his time between educating and creating. He holds interests and owns several creative projects that entertain, enlighten, enhance, and educate, hoping to inspire and motivate you.

Follow, find new works, and stay up to date
with Walter the Educator™
at WaltertheEducator.com

www.ingramcontent.com/pod-product-compliance
Lightning Source LLC
LaVergne TN
LVHW051958060526
838201LV00059B/3711